AF890438

Easy as Line AB and Point C

Written by Christopher Becker
Illustrated by Deanna Becker

Published by Gamify Education LLC
100 N. Howard Street, Suite 4791, Spokane, Washington 99201

Copyright © 2025 Gamify Education LLC

All rights reserved. This publication may not be copied, photocopied, reproduced, translated, or converted to any electronic or machine-readable form in whole or in part, except for brief quotations, without prior written approval from Gamify Education LLC.

ISBN: 978-1-962239-02-8

To every hard-working math teacher out there who has struggled to find time in the school year to teach geometry...

...this book is for you.

How rude of me! I haven't introduced myself.

I am a point. 

I live in space, just like you. But unlike you, I don't take up any space at all!

I am quite special.

My name?

Well, I am called many things.

One of my popular names is A, as in point A.

You can call me that.

One thing you should know about me is that I can "change" sizes.

This is me:

And so is this:

And even this:

Did I actually get bigger or smaller?

No, it just looks like I did. I told you I was special.

This is my best friend: 

You can call him B, as in point B, but he goes by lots of other names, too. He's just like me!

We like to get together with our friends. We connect a lot.

When we do this, we form something brand new.

We become a...

Line Segment

But not just any ordinary line segment...

Point B and I are really important when we connect like this.

One of us is first, and the other is last.

It doesn't matter who is first in line. Either way, we stay connected.

You can call us line segment AB. Or, you can call us line segment BA.

We like both names.

Meet another one of my friends: 

She likes to go by C, as in point C. She wanted to join in on the fun.

"Welcome to our line segment, C!"

Point C wanted to be at an end.

Do you know what I told her?

I said she already was!

Point C was at the end of one line segment with me...

...and at the end of another line segment with B.

Because C came along, she helped us create three line segments of different sizes.

That makes her pretty special.

Just as I got done explaining this, a bunch of friends started lining up next to point B.

And guess what?

They are STILL lining up next to him!

Point B wasn't sure if we were a line segment anymore.

I told him we still were...

...but we also formed something new.

Because of the friends who lined up next to B, we became a...

Ray

But not just any ordinary ray...

Point B thought we should be called ray AB.

Point C thought we should be called ray AC.

I said that we could be called both names.

C asked if she could start the ray...

Point C could start a ray of her own, but it wouldn't be as long as the ray I started.

Point B wondered why he couldn't start his own ray.

I said that he could—but first, he would need to know the name of a friend BEHIND him.

Just as Point B was figuring that out, a bunch more friends started lining up next to me.

I can't even tell you how many friends showed up!

But because they did, we formed something else.

We became a...

Line

But not just any ordinary line...

Point B and I are on the line together, so we can be called line AB or line BA.

Point C is on this line too, so we can be called line AC or line CA.

That's when B shouted...

"STOP EVERYTHING!"

Point B said we were much shorter without him, but that isn't true.

I told him to think about all of the friends next to point C.

The line with C is just as long as the line with B.

After connecting these dots, B figured out our other two names.

That's when we all heard a...

CRASH

...and then, off in the distance, another crash.

I looked around and asked, "Is everyone all right?"

Point B said he was fine but mentioned that it felt like something went right through him.

And sure enough, something did...

Meet two more friends of ours, point D and point E.

Point D told us that they were just passing through.

And because they did, something changed.

Together, we became...

Intersecting Lines

Point B cheered because he now belonged to two lines.

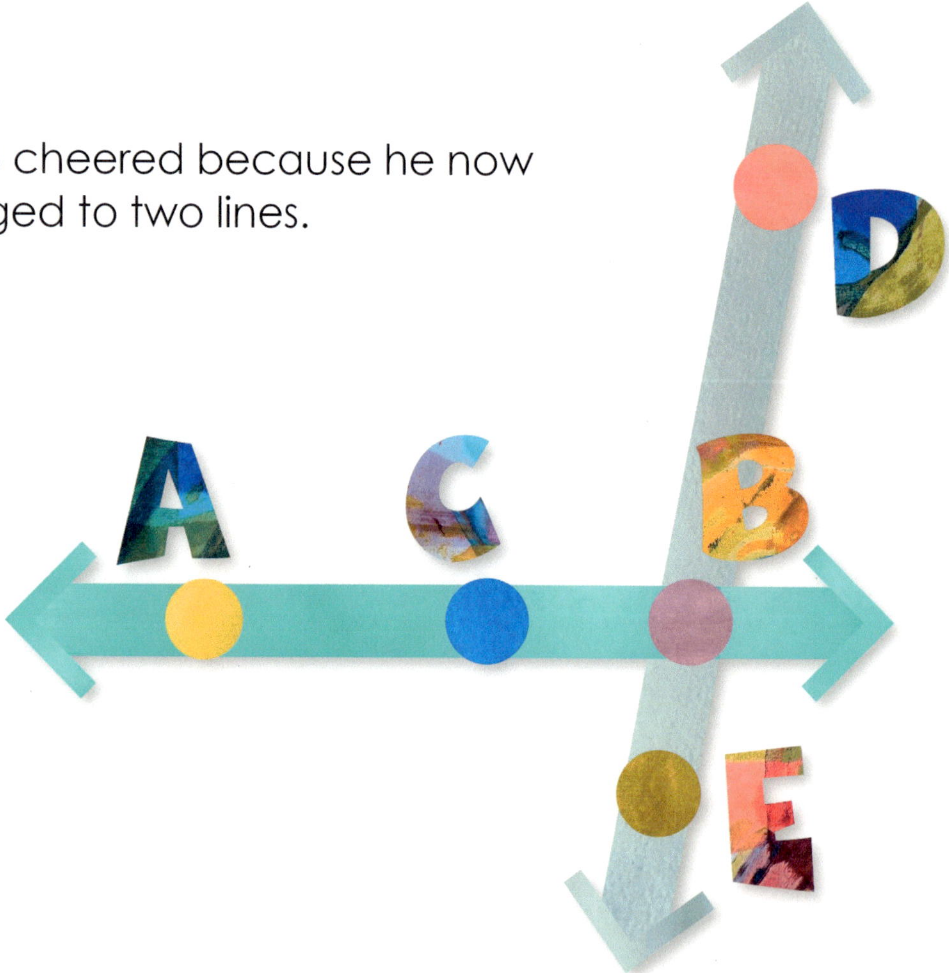

But the celebration didn't last long because there was another...

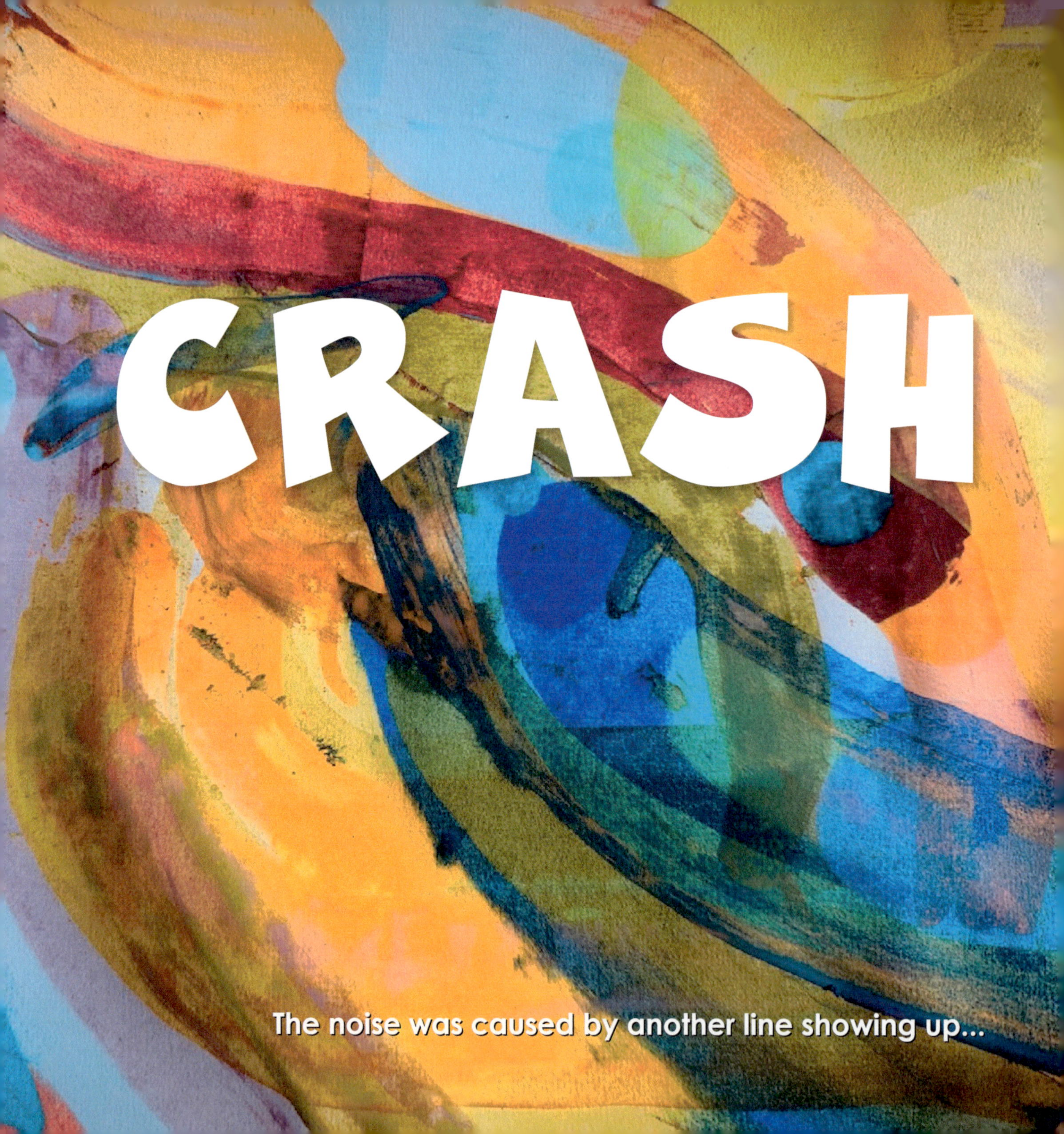

CRASH

The noise was caused by another line showing up...

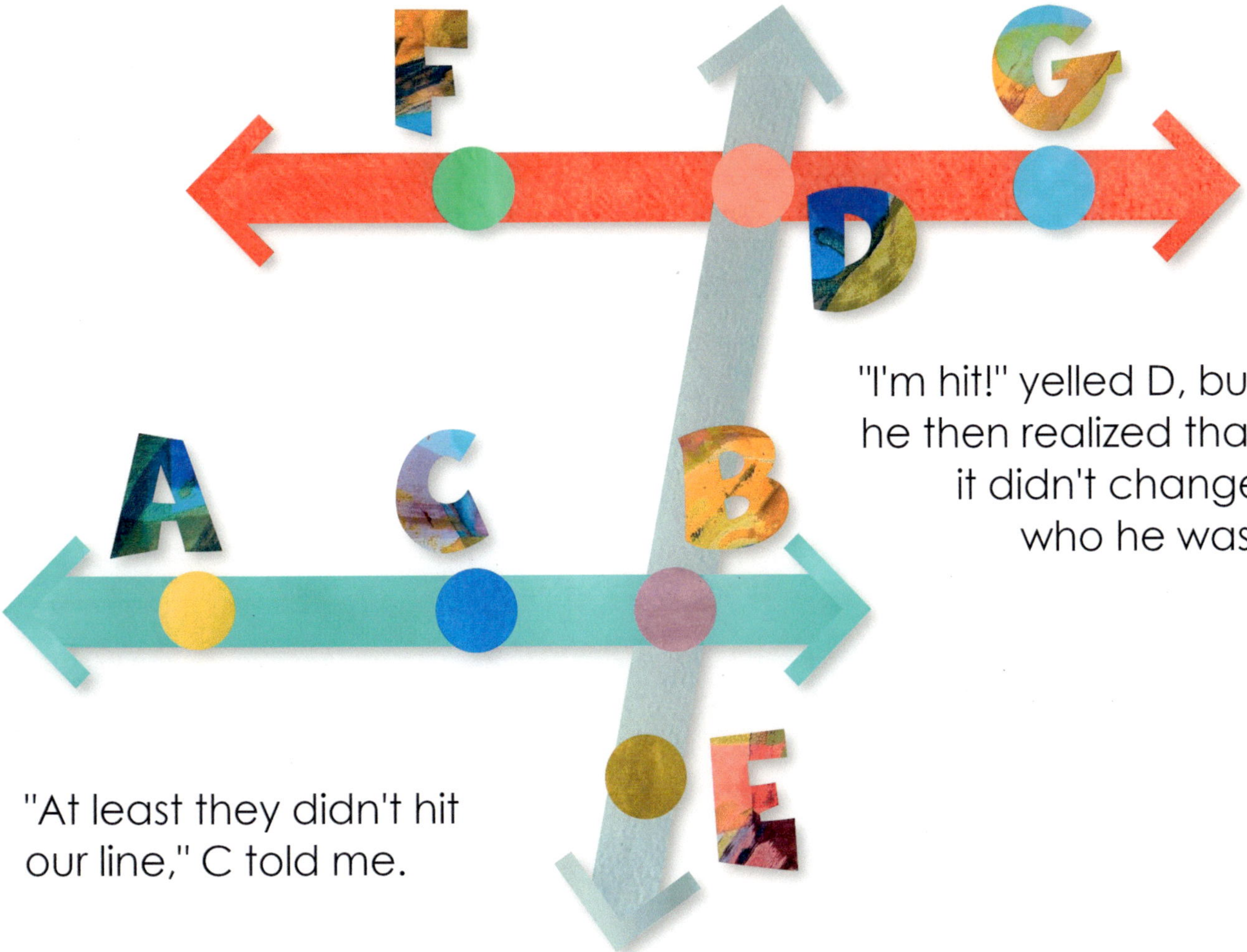

"I'm hit!" yelled D, but he then realized that it didn't change who he was.

"At least they didn't hit our line," C told me.

I assured her, "And they never will."

Because of line FG, we got another new name.

Together, we became...

Parallel Lines

See? We can't possibly ever cross each other.

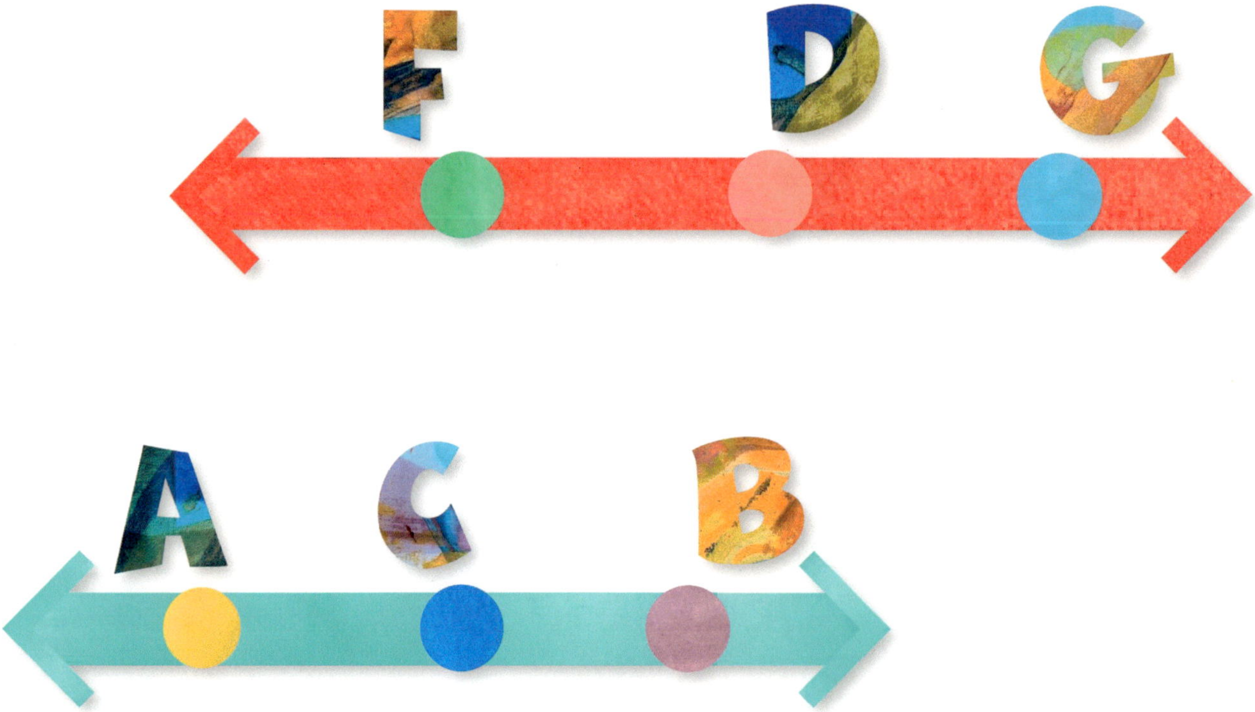

At this point, we were all ready for some peace and quiet.

But instead, we heard a...

CRASH

And then another...

CRASH

And still, another...

CRASH

The points on line HI all said hello as they crossed F and C.

But something about this line was different. It seemed to stand so tall compared to the line B and I were on.

And with that, we got one last name as we became...

Perpendicular Lines

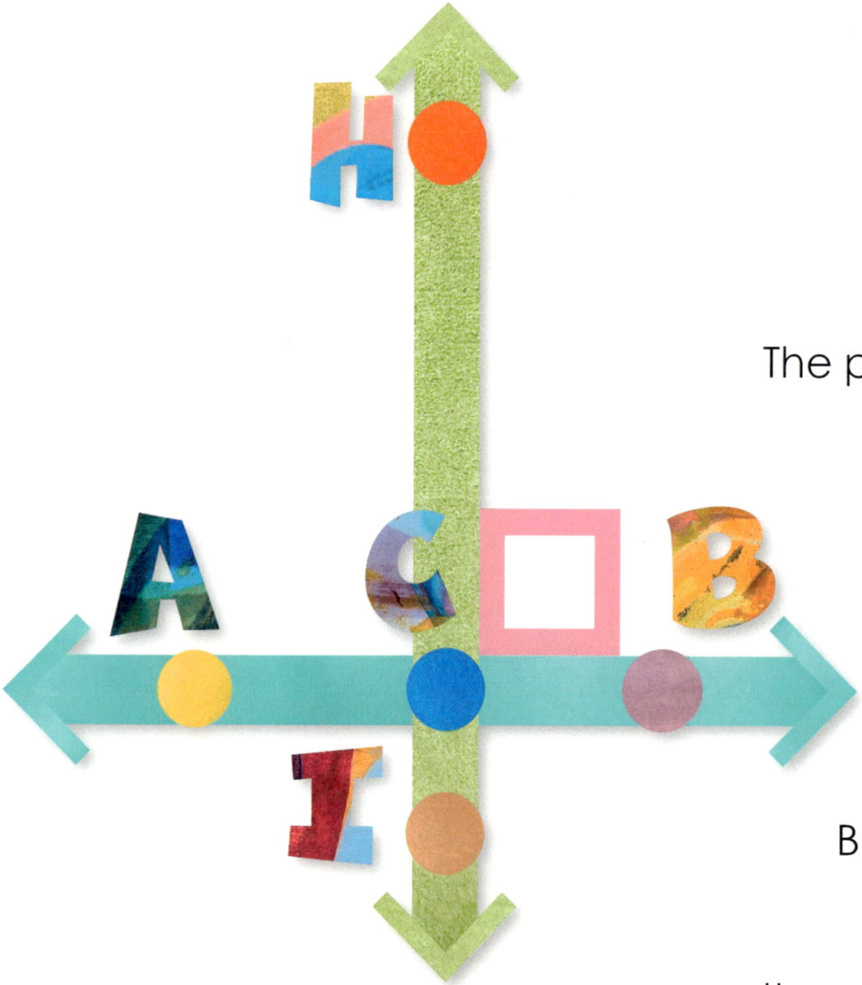

The points on line HI met us
at just the right angle.

But why three crashes?

Only two points said
they were passed through.

While everyone was trying to figure out this mystery, that's when we heard...

...a soft voice from a friend whose name we didn't even know.

She pointed out that line HI went through her as well.

We all welcomed her to our space, but at the same time, we apologized for not even noticing she was there.

So there you have it—our journey together seems to have come to an end.

It all starts with friends like me. 

From there, we make line segments, rays, and all sorts of lines.

Thank you for spending this time with us.

It was so nice to meet you.

But this can't really be the end, can it?

Wait...who are you?

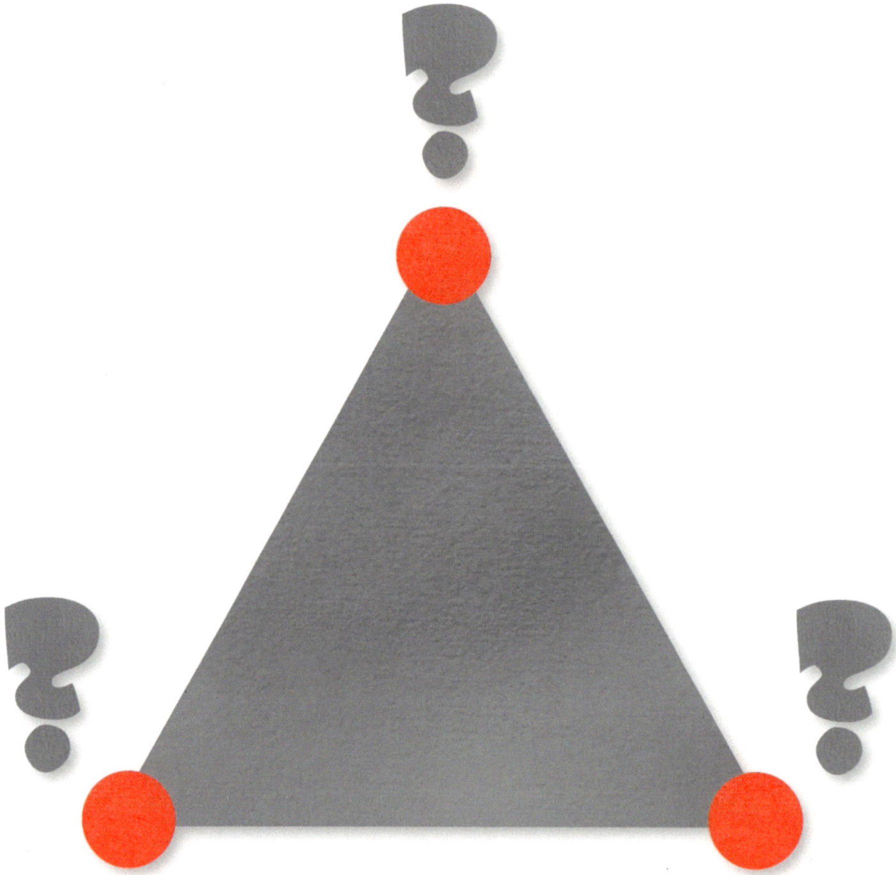

To be continued...

continue reading for some deep mathematical thinking

Points are non-dimensional, meaning they have no length, width, or depth. Since you can't truly draw something without dimensions, you can represent points as tiny circles. It's the best you can do.

You might think you can count all the points on a line segment, but you can't. Even the shortest segment contains an infinite number of points. This is one of the reasons non-dimensional objects are so special.

Remember when point A said that the points are still lining up next to Point B?

That's because rays extend infinitely in one direction. No matter how long you draw them, they never really end.

This also explains why Point A didn't know how many friends lined up next to him—lines extend forever in both directions.

To represent infinity, use an arrow for a ray's one-way extension and two arrows for a line.

Even though he wanted to, Point B couldn't start a ray of his own.

Why not?

When naming a segment, ray, or line using points, you always need TWO of them.

To name a ray, you need a second point to show the direction it extends toward infinity.

That's why point A told B that he would need to know the name of a friend BEHIND him.

Point B thought a shorter line was actually shorter—but in reality, all lines are infinitely long.

When you draw a line, remember that what you see is only a small part of a much greater whole.

What was going on with that unidentified point at the end?

Even if two lines look like they don't intersect, if they are on the same plane and aren't parallel, they must intersect somewhere.

The exact location may not be drawn, but mathematically, it exists—at some point.

Did you catch it?

A reference to the unidentified point appeared much earlier in the story:

The first crash the points heard was caused by line DE intersecting point B. But that distant crash? That was line DE intersecting the unidentified point—long before anyone realized it was there.

And the cliffhanger?

Will these points go on to form angles and triangles?

You will have to wait and see.

About the Author:

Christopher is the big brother.

About the Illustrator:

Deanna is the little sister.